The expression "a picture is worth a thousand words" is particularly appropriate when it refers to medical photographs. Words alone can never create the profound impact of photographs that vividly illustrate diseases a physician might still encounter, as well as those that have already been brought under control.

Today, we no longer consider epidemics of the past as public threats. Yet, with ever-changing global dynamics, physicians must be prepared for the possible return of past scourges. An example is smallpox - shocking but possibly real because of today's new and frightening threats. President Bush recently announced that to protect Americans, the smallpox vaccine will be made available, beginning with the military and health care workers. But preparation begins in many ways, including an understanding and recognition of the physical, mental, and social impact these diseases once had on the population. Medical photography can help us reach that understanding.

GlaxoSmithKline, in its continual effort to serve the public welfare and the medical community, has commissioned an original educational series of four photographic books on the history of medicine from the Burns Archive of Medical Photographic History. These original works, titled "Respiratory Disease: A Photographic History 1845 to 1945," will encourage physicians to think about past medical practices and how medicine has progressed over its most critical century.

This collection, by renowned physician and historian Stanley B. Burns, dominates the field of early medical photography, containing more than 50,000 medically significant photographs. Many of them, complete with written explanations, have never been seen by the general medical profession.

Dr. Burns has published more than a dozen books and his collection has been the subject of numerous exhibitions. Recent presentations have been mounted at the Musee d'Orsay, Paris; Kulturbro 2002 Art Biennial, Brosarp, Sweden; The Center for The Study of the United States, Haifa, Israel; and the National Arts Club, New York. Dr. Burns' full collection of over 700,000 photographs are used by researchers, book publishers, media and film companies worldwide.

GSK is proud to sponsor this original Historical Medical Educational Series.

Chris Viehbacher
President, US Pharmaceuticals
GlaxoSmithKline

RESPIRATORY DISEASE:

A PHOTOGRAPHIC HISTORY
1896-1920 THE X-RAY ERA
SELECTIONS FROM *THE BURNS ARCHIVE*

STANLEY B. BURNS, M.D.

BURNS ARCHIVE PRESS
NEW YORK 2003

Colophon

This first edition of *Respiratory Disease: A Photographic History, 1896-1920 The X-Ray Era* is limited to 20,150 copies including a special cased edition of 1000 copies. The photographs are copyright Stanley B. Burns, MD & The Burns Archive. The design is copyright Elizabeth A. Burns & The Burns Archive. The text and contents of this volume are copyright Stanley B. Burns, MD, 2003. Printed and bound in China for The Burns Archive Press, NY, a division of Burns Archive Photographic Distributors, Ltd. NY. The book is printed on acid free 140 GMS Hi-Q Matte paper. The 4-color separations were scanned at 200 lines per square inch.

ISBN 0-9612958-7-4

Library of Congress Cataloging-in-Publication Data:
Burns, Stanley B.
Respiratory Disease: A Photographic History: 1896-1920 The X-Ray Era
    Includes bibliographical references: 1. Medical, History 2. Respiratory Disease
    3. Photography, History 4. World War I 5. Lung Disease 6. Infectious Disease
    7. Influenza 8. Stanley B. Burns, MD

The Burns Archive Press

Author & Publisher: Stanley B. Burns, MD
Editor: Sara Cleary-Burns
Production & Design: Elizabeth A. Burns

Photographic Captions

Front Cover: Therapeutic Pneumothorax Treatment, 1915
In 1915, Dr. Kennon Dunham published a series of stereoroentenograms of lung disease and its treatment. (see Photo 13). In this case, Dr. Dunham demonstrated the use of nitrogen gas injected into the intra-pleural space to collapse the lung and stop repeated hemorrhaging due to tuberculosis. "Right Lung Completely collapsed by nitrogen gas." This view shows the greatest extent to which a lung can be collapsed. The lung is held by pleural adhesions to the lateral wall. "The hemorrhage has been completely controlled after the second injection (of gas)."

Frontispiece: Doctor with Stereoscopic Machine, 1914
A physician looks into the viewer of an innovative device to allow three dimensional viewing of x-rays. Dr. Kennon Dunham of Cincinnati promoted the use of stereoroentgenograms as an aid to more accurate diagnosis of chest disease. Analysis of x-rays in three dimensions has remained an important part of radiology. The CAT, MRI, and PET scans provide modern methods of three dimensional analysis of the body.

Back Cover: Artificial Respiration Position for Patients on the Operating Table, 1898
Respiratory arrest or aspiration of stomach contents during surgical anesthesia has been a feared complication since its discovery in 1846. In this late 1890s view an innovative technique developed by anesthesiologists at the Johns Hopkins Hospital is used to revive a patient. The surgical assistant jumps on the operating table and raises the patient to a head down position which would help empty the lungs of aspirated contents. Abdominal contents though compromising the lung volume would help in the expulsion of bronchial aspirate. This problem was solved by the development of the cuffed endotracheal tube that permitted control of the airway.

Contents

PREFACE

As an ophthalmologist, a life-long collector and a historian, I am fascinated by our past and drawn to visualizing history. When I first started collecting medical photographs in 1975, I chose images based on both their importance as historic documents and as evidence of medicine's rich past. What I soon realized was their artistic strength.

In 1979, I created the Burns Archive, which is dedicated to preserving medical photographs and producing publications on the history of medical photography. By the mid-1980s, noted curators and artists became interested in medical photography as art. Marvin Heiferman curated "In the Picture of Health" in 1984, an exhibit of more than 140 photographs from the Archive. This was the first exhibition of medical photographs in a public art institution. In 1987, Joel-Peter Witkin edited *Masterpieces of Photography: Selections from the Burns Archive*. Since then, numerous major museums and galleries, recognizing the artistic value of these images, have started to collect and exhibit medical photography. These institutions now display vintage medical photographs of patients, procedures and practitioners to the general public.

What I have learned these past twenty-eight years is that art matters. Art elevates and stimulates us to see things differently. Art creates a different perspective and point of view. When medical photographs are presented to the public, the images are viewed and conceived in terms of personal mortality, human fragility and the vagaries of life. Terror and fascination draw the non-medical public into dialogue with these images. Although the art world's appreciation of vintage medical photography as art is laudable, my original goal as a historian was to present these photographs to my colleagues not as art, but as historic documents. I want my fellow physicians to visually experience the practice of medicine in the 19th century to help them gain a better understanding of the foundation of our therapies and patient treatment. As a physician, I am brought back to a different reality by these photographs. I see my patients; I see difficulties in therapy, I see personal challenges and I wonder whether what I am doing and what I believe in will one day be proven wrong.

GlaxoSmithKline has given me the opportunity to share with my colleagues these photographs on respiratory disease. This compilation is not meant to be an encyclopedic history of the topic, but to put the emphasis on artistic, medical photographs that will allow you to see the transition of medicine from yesterday to today. Many of the treatments depicted have long since become outmoded, but our predecessors believed they were offering the best therapy available. My hope is that you will look at these images as icons of our past and gain a better understanding of what we do and how we can better serve our patients.

Medicine's quest to unselfishly help and heal is one of mankind's highest goals. I am proud to be part of the medical profession and share these photographs to further that goal.

Stanley B. Burns, M.D., F.A.C.S.
New York, March 2003

THE X-RAY ERA 1896-1920

The years 1896 to 1920 were unprecedented in the history of the world. It was a time of miracles. Not only in medicine but in the development of consumer products. The airplane, automobile, movies, telephone, radio, and a host of other devices were invented or further refined that allowed the working and middle classes pleasures and potentials, which were to define the twentieth century. Electricity had not only brought the x-ray but new electrical products from the electric chair to electric lighting.

1896 was a landmark year in medical history. The x-ray had just been discovered and this new diagnostic tool would usher in a new age in the treatment of respiratory tract disease. Physicians could now evaluate the internal structures of the body. Pioneer radiologists and surgeons developed contrast material, new procedures and invasive technique. In this volume, we document some of the complications of tuberculosis and the innovative procedures turn of the century physicians used in their attempt to help and heal.

The devastation created by trench warfare and the new weapons of World War I particularly affected the head and respiratory tract. Poison gas and jellied gasoline flame throwers induced serious respiratory damage and death. Trench warfare caused massive numbers of head and neck wounds as that was the body part most often exposed to shrapnel and snipers. Maxillofacial, head and neck and plastic surgeons labored to piece back together faces and parts of the upper respiratory tract from these wounds.

During the first decades of the twentieth century the x-ray and the laboratory resulted in an enormous increase in diagnostic skills and propelled hospital-based surgery to the forefront of medical achievement. The public image of the hospital changed from a place to die to a symbol of health restoration. Photographs of the operating rooms and hospital facilities became an important part of projecting that image. Technical daring and improvements in instruments allowed innovative surgeons, guided by x-rays, to develop sophisticated procedures. Unmitigated technical boldness and advances in local and general anesthesia allowed physicians, surgeons and otorhinolaryngologists to develop new or modify existing procedures that tackled diseases of the chest, sinus, ear, mastoid and throat. New treatment modalities such as Radium for cancers, light therapies for tuberculosis and skin diseases and the identification of nutritional and endocrine diseases created new specialties. In surgery the specialties of urology, neurosurgery, maxillofacial surgery were firmly established.

Endoscopy flourished in this era. The bronchoesophascope and other tools to look inside the body were perfected. Coupled with the x-ray it seemed no part of the human body was invisible or inaccessible. The lungs and sinuses, cauldrons of infections, were now visible by one means or another making accurate diagnosis and therapy possible. Several centuries of old scourges, malaria, typhoid, typhus, and yellow fever were conquered or controlled. The specific therapies and vaccines developed in the first two decades of the twentieth century led the public to believe that the conquest of disease was imminent. Two new epidemics appeared in this time to distort medicine's achievements. Both killed by attacking the respiratory tract. Influenza and poliomyelitis.

1
PERCUSSION: FIRST PROCEDURE FOR DIAGNOSING CHEST DISEASE
YVON, PHOTOGRAPHER
PARIS
CIRCA 1920

Percussion has been a critical part of chest disease diagnosis since the early nineteenth century. This physician in a French chest clinic is percussing (tapping) the patient's back to evaluate by the resonance the presence or absence of fluid or other mass. The first practical diagnostic tool in examining the chest - percussion was developed by Viennese physician, Leopold Auenbrugger, M.D. (1722-1809). In his 1761 text *Inventum Novum Ex Percussione* he noted the sound of tapping on the chest "produces analogous results to those observed by striking a cask, for example, in different degrees of emptiness or fullness." Auenbrugger verified his findings with postmortem examinations and experimentation with cadavers. He filled their lungs with various amounts of fluid and evaluated the sounds produced. Despite the translation of his text into French by Roziere de la Chassagne, M.D. in 1770, his exemplary research did not gain immediate favor. The value of percussion finally was heralded in the 1808 classic by Jean Nicholas Corvisart, M.D. (1755-1820). With Dr. Rene Laënnec's invention of the stethoscope in 1818 and his publication of two volumes on chest disease, 1819 and 1826, the main diagnostic tools for diagnosis of chest disease were established.

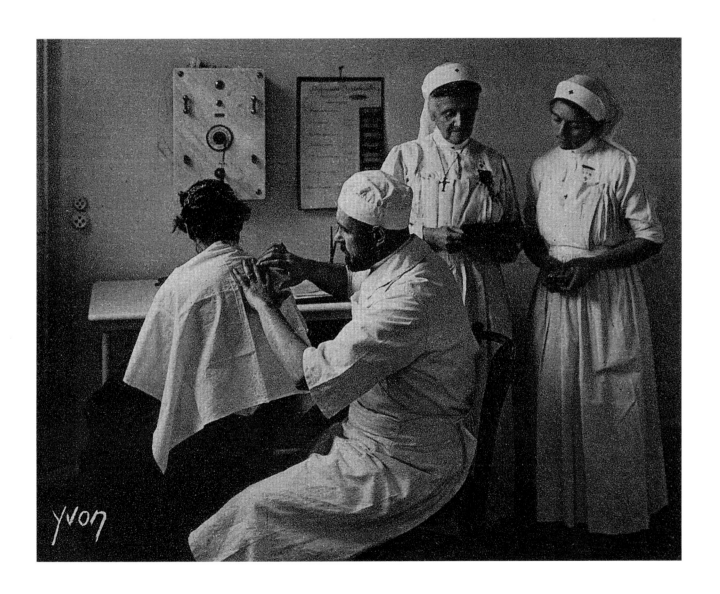

2
REMOVING PUS FROM A LUNG ABSCESS DUE TO PNEUMONIA
1912

In the pre-antibiotic era, bacterial infections of the lung often resulted in cavity formation and abscess. Tuberculous empyemia and other causes of empyemia were often treated in the same manner. In the mid-nineteenth century physicians recognized that tuberculosis cavities would form with pleural adhesions, spontaneously develop fistulas which then drained the contents. Some postulated it was possible to make an incision in the abscess area and drain it without causing pneumothorax. Pulmonotomy, simple drainage of a lung cavity, not only evacuates its contents but allows introduction of disinfectants. This treatment would hopefully destroy the bacteria and speed healing. In the 1890s, surgeons performed this procedure and used strong antiseptic solutions, even chlorine gas, to destroy the bacteria. Though chlorine had been a popular therapeutic gas among gas therapists, pulmonologists, since the 1790s the results were not generally favorable. Finally in 1895, one Parisian, thoracic surgeon, Paul Reclus, M.D. (1874-1914), summed up the results of the procedure: "In spite of our first enthusiasm, we must interfere only in exceptional cases; and even if the cavity or the trouble that arises from it can be lessened, the original cause of it, the tubercular disease itself, is still left behind, past all help from incision and drainage." The procedure fell into general disfavor and thoracic surgeons turned their attention to artificial pneumothorax to treat pulmonary tuberculosis. Thoracoplasty and other chest operations gained support as modern surgical techniques improved. This patient has just endured the drainage of a chronic lung abscess and the nurse is suctioning the pus. A dressing impregnated with a disinfectant will be applied to the wound completing the treatment. Since open bacterial wounds were commonplace it's no wonder tuberculosis and other infectious diseases rapidly spread.

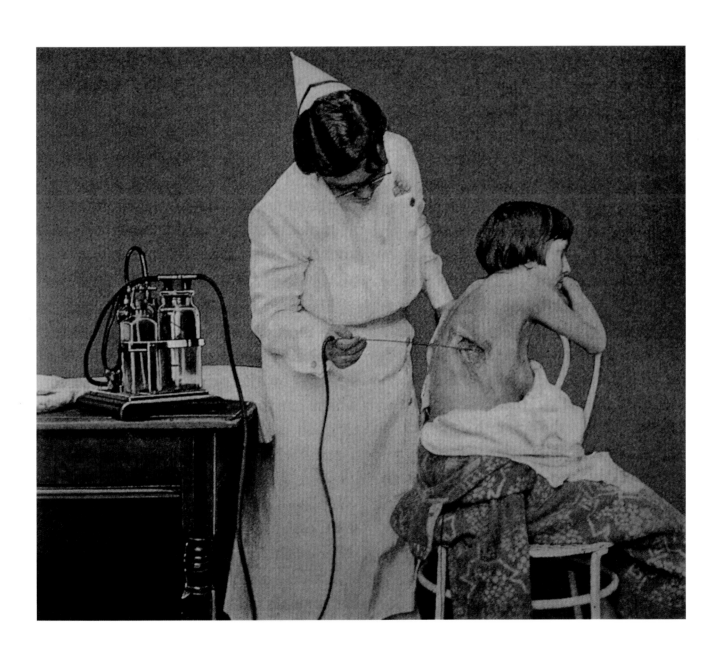

3
Nurses with an Incubator Respiratory: Assistance for the Premature Infant

E. Belinfaute, Photographer
Philadelphia
May 12, 1897

Underdeveloped lungs with concomitant respiratory distress is among the serious problems a premature infant faces and one of the leading causes of their death. One of the marvels at the turn of the century was the invention of the incubator by Marx of New York. The incubator was used to treat and nurture premature infants delivering warm air to a vented closed heated container. The simple warmth helped babies survive. Although today younger and younger infants are surviving because of the care received in the modern neo-natal units, respiratory function remains one of the major hurdles. Before the introduction of the incubator premature labor, whether started by natural causes or induced due to a medical emergency such as pre-eclampsia or placenta previa, often resulted in a fatal outcome. With the new device it was touted that 85% born at the 36th week of gestation (ninth lunar month) could be saved if placed in an incubator, warmed and force-fed. The Paris Maternity Hospital at the turn of the century reported the incubator saved 64% born at 7 months and an astonishing 30% survived at sixth months. As late as the 1960s a baby less than five pounds was considered a preemie. Its survival was questionable. Today babies weighing as little as a pound are being saved from death.

The incubator was inexpensive and easy to construct. The original model by Marx was a wooden box divided into two compartments, lined with asbestos and zinc. One compartment held the baby, the other held a mechanism used to accept steam from a copper boiler. A glass sliding cover or door allowed observation of both the baby and a thermometer and numerous vent holes provided air circulation. A well-padded basket held the baby suspended above the steam radiator/heating device. While it was all extremely easy and cheap to manufacture its use required round-the-clock attendants. Viewed in that light, it was a pioneer model of the infant intensive care unit. This photograph was taken at the Preston Retreat, May 12, 1897. Three nurses are operating an incubator. The design is a variation on Marx's device. One nurse is testing the quality of the air by inhaling a sample through a tube attached to the apparatus. The marvels of modern medicine always draw the curious and the incubator became a public spectacle. For decades infants in incubators were displayed at world fairs and expositions. By 1910 they had even become attractions in sideshows. Coney Island boasted such an exhibit up to World War II.

The incubator opened a Pandora's box in the treatment of premature and diseased infants. The advance of medicine in the past century can easily be measured by the progress made in premature infant care. Treatment has become increasingly sophisticated until now there are complete intensive neo-natal care units with numerous neo-natal specialists. Technology has advanced to the point where hand held blood gas analyzers use a drop of blood and furnish results in less than a minute. The treatment of preemies has evolved from the inexpensive incubator to, perhaps, the most costly and labor intensive of all medical endeavors with the result that many survive with anomalies and disease states that would have previously destined them for an early death.

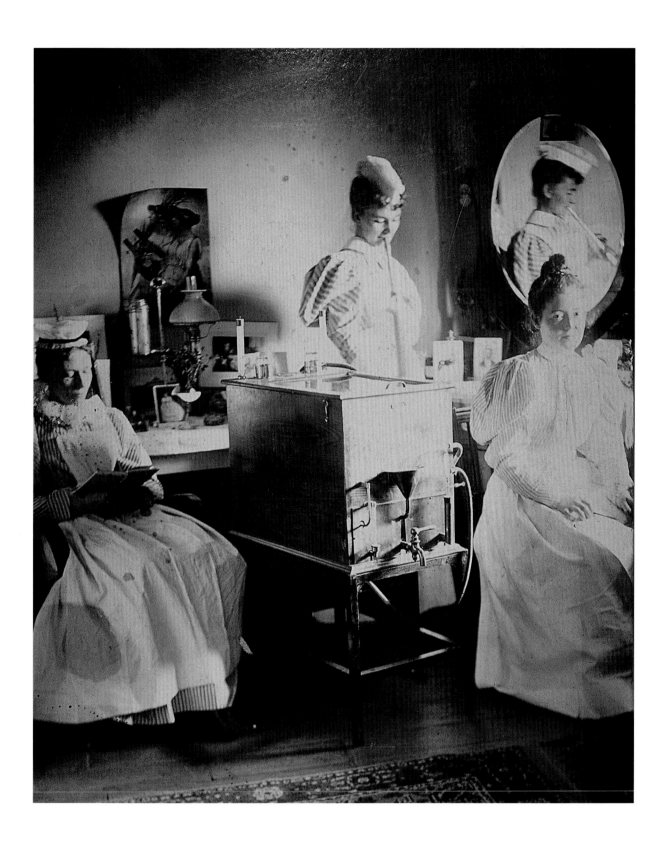

4
TREATMENT FOR LUNG ABSCESS BY INJECTING CHEST CAVITY WITH BISMUTH

STEREOVIEW
CHICAGO
1908

A major advance in treating lung disease was posited in the late 1840s by pioneer New York throat specialist, Horace Green, M.D. (1801-1866). Green suggested medications should be injected directly into the lung. Green reported he had passed a 12 1/2 inch tube, saturated with a solution of silver nitrate into each bronchus. In 1846 this data was published in his book, *Treatise on Diseases of the Air Passages*. The report was not believed and he was compelled to resign from his medical society. It was not until the 1870s that Dr. Green was vindicated. A thoracic surgeon, Mosler, injected various agents into the lung and reported the successful results. In 1882, Fraenkel, after testing numerous agents, also noted that the human lung could accept antiseptic agents. Slowly thoracic surgeons began the practice of intrapulmonic injection therapy. By the first decade of the twentieth century they had successfully developed substances and surgical procedures that would revolutionize the treatment of lung disease. One such proponent, Emil G. Beck, M.D., Surgeon to the North Chicago Hospital in 1906 developed the technique of bismuth paste injection into the lung. He had perfected the technique after learning of the special contrast properties of the substance. Walter B. Cannon, M.D. (1871-1945), had used bismuth as a contrast media in 1898. In 1911, Beck published his findings illustrated with original tipped-in, stereoview photographs of his technique in the noted *Stereo-Clinic* book series edited by Howard Kelly, M.D. of Johns Hopkins Hospital. These photographs are from Beck's *Diagnosis and Treatment of Suppurative Sinuses and Empyema With the Aid of Bismuth Paste*. Bismuth paste was a toxic substance that could poison the patient if excessive absorption occurred. Beck advised "the paste should be removed immediately by filling the sinuses or cavities with warm, sterile olive oil and withdrawing the mixture within twenty-four hours by means of suction." Beck's case report reads as follows: "This 15 year old patient after swallowing a brass tack developed pneumonia. A lung abscess developed." In 1908, Beck evaluated the patient, operated and applied his treatment. Under the guidance of stereo-roentograms he resected four ribs and opened the abscess. He then filled the wound with gauze packing and put the patient to bed. On the third day, under chloroform anesthesia he removed the gauze and opened the abscess. About 100 grams of very offensive pus escaped. The following day "the cavity was injected with 100 grams of 33 per cent Bismuth-vaselin paste. (see photo bottom left). It is not advisable to inject the paste while the patient is anaesthetized since it may run into the opposite bronchus and thus suffocate him." Dr Beck presented the case to the Chicago Medical Society. The patient did well, moved to California where "one year later he suddenly developed a fever" He returned to Chicago where x-rays showed the tack was still present and causing a new abscess to form. The patient was operated upon, the tack and abscess removed and treated as before with bismuth injections. The final photograph shows the patient completely healed. Dr. Beck's technique became a standard procedure for treating lung infections as well as tuberculous fistulas and abscess in various parts of the body.

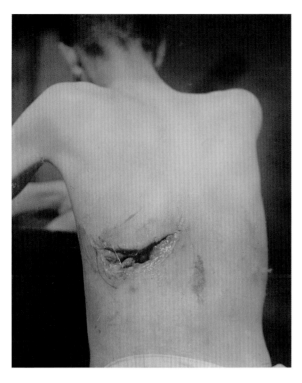

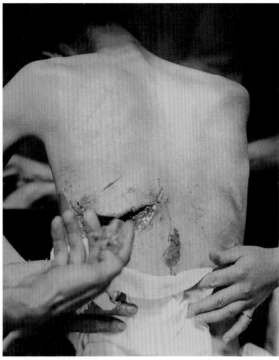

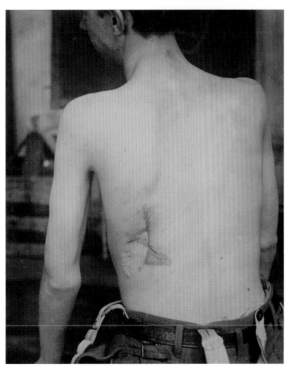

5
LISTENING TO THE CHEST WITH MONAURAL STETHOSCOPE
FRANCE
CIRCA 1919

This is a classic pose of a physician listening to a patient's chest with a monaural stethoscope. He is demonstrating the proper use and position of the instrument. European physicians used and posed with monaural stethoscopes until the 1940s. The monaural stethoscope, invented in 1818 by French physician Rene Laënnec remained the standard instrument for examining the chest in much of the world, because of European influences. In 1855, an American, Dr. G. Cammann, produced a practical and superior instrument, the binaural stethoscope with flexible rubber tubing. The binaural scope not only offered better acoustics but ambient sound was drowned out because both ears were used. Another major advantage is illustrated in this photograph. Not only did the short, about 7 inch, monaural instrument require a physician to get close to a patient who often had severe, contagious, infectious disease but the instrument also had to be placed squarely on the skin, again putting the physician uncomfortably close. The long rubber tubes of the binaural stethoscope allowed the physician to listen to the chest at a safer distance.

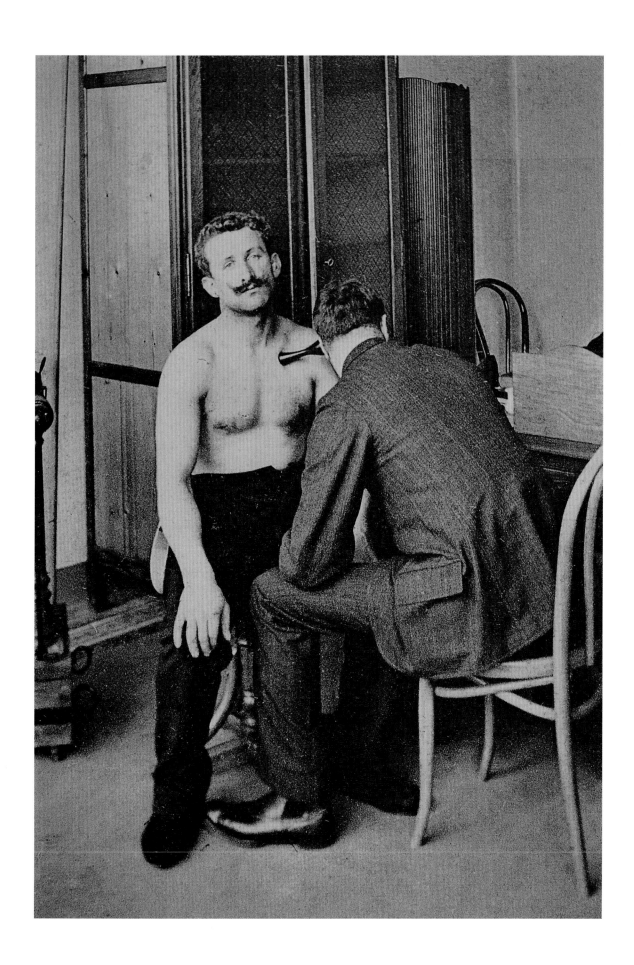

6
NEGATIVE PRESSURE FOR RE-EXPANSION OF THE LUNG
STEREOVIEW
CHICAGO
JANUARY 1910

Tuberculosis and other infectious diseases of the lung often resulted in various conditions that would cause the lung tissue to collapse or contract. In the first two decades of the twentieth century, Dr. Emil G. Beck, a Chicago surgeon, experimented with various procedures to ameliorate a number of these complications. Providentially Beck photographed both his patients and technique providing an important legacy to physicians by documenting the dedication and innovative work they perform in their attempt to help, heal and conquer disease. In the pre-antibiotic era tuberculosis was the leading cause of death and had been for over for two centuries. Almost every substance and manipulation was employed in the treatment of this disease. The devastation to daily life and health tuberculosis patients endured is reflected in this photograph. Unlike cancer that can kill fairly quickly, this disease can eat away for decades. At the suggestion of his brother, Dr. Carl Beck, Emil Beck developed a procedure for re-expanding lung tissue using negative pressure. This is his description of this case:

Mr. M., thirty-one years old, developed acute pleurisy, followed by empyema. After several tappings, permanent drainage with rubber tube was established. The empyema proved to be of tubercular origin, the bacilli having been constantly found in the pus from the pleural cavity. During the past nine years the cavity has been flushed daily with antiseptic solutions and drainage kept up. On January 10, 1910 the first bismuth paste injection was made. The cavity was not filled entirely but 240 grams were injected. The discharge soon became sterile and serous, but the expansion of the lung was very tardy, and therefore this new method by means of negative pressure was employed with satisfactory results. The rubber tube, which is fastened to a nipple, is inserted into the cavity, and is covered by a Bier's cup (see photo). The connecting attachment of a large suction syringe, which has a release valve, is attached to the outlet of the Bier's cup. Moderate suction is produced and released in rhythm of the patients breathing. During inspiration we produce suction; during expiration we release the valve and allow the lungs to collapse. This treatment is carried on systematically every day for five minutes. This (procedure) is suitable in cases in which the discharge is not profuse or bloody. In some cases the granulating cavity is very susceptible to oozing and this might be very much aggravated by strong suction (the negative pressure). In this case we have noted a gradual expansion of the lung, and, while the cavity held 240 grams of paste at first, it now overflows when injected with 15 grams, and the physical signs corroborate the fact of lung expansion.

In 1911 Dr. Beck published this and other interesting cases in the *Stereo-Clinic,* a series of texts using photographic stereoviews as illustrations for physician education. They were the equivalent of the modern video used today.

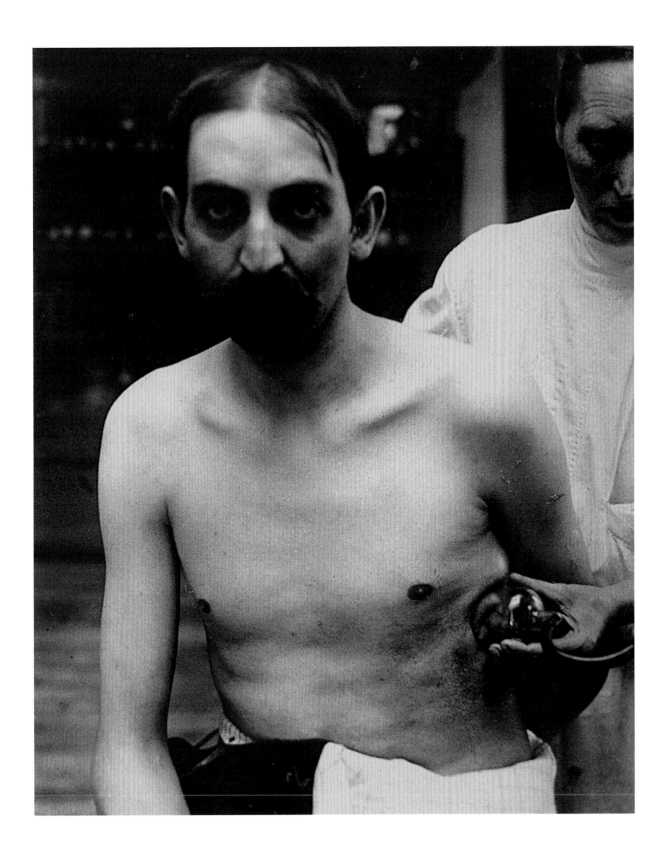

7
Dr Haffkine & Staff, Bubonic Plague Specialists
Harbin, China Plague Hospital
Chinese Epidemic: 40,000 died in 3 1/2 months
1911

8
Four Types of Carts for Chinese Plague Victims
Stretcher for Emergencies, Clothes Cart, Ambulance Cart & Death Cart
Fu Chia Dien, China
1911

9
Burning Plague Houses
Harbin, China Plague Epidemic
Fu Chia Dien, China
1911

These photographs are from a series of ninety-six images from a rare album documenting the 1911 outbreak of plague in Northern China. Here, Dr. Waldemar Mordecai Wolff Haffkine, MD (1860-1930), one of the revered figures in vaccino-therapy together with American and Chinese doctors cared for the victims. In the album, many aspects of the epidemic from sanitary precautions, serum preparation, to autopsies are depicted. The plague was carried by flea-infested rats. At times, rats were considered food. In ancient times the Chinese fondly called them "house deer." Waves of devastating plague epidemics have ravaged China since the 14th century. In 1911, a newly constructed railway in Manchuria and Northern China was the route of a deadly plague. Over 600,000 people died within a few months. Several other epidemics followed not only in China but in India and Southeast Asia as well. It is estimated that 13,000,000 died. Ultimately better sanitation, efforts in rat killing, antibiotics, and vaccines have controlled the disease and, in 1959, the World Health Organization declared the two century old, third pandemic of plague to be over. However, until the late 1970s, there were over 100,000 reported cases of plague a year. It still occurs with small sporadic, isolated cases acquired from wild rodents. About a dozen people die each year in the Western United States from plague carried by rodents.

During the 1911 epidemic China appealed outside their borders for medical assistance for the very first time. A world conference of foreign medical specialists convened in Manchuria to deal with the plague. This series of photographs record the work of Dr Haffkine. He developed and introduced prophylactic immunization for two of mankind's dreaded epidemic killers, cholera and plague. Assistant to Pasteur until 1893, he was invited by the government of India to work at a new research center for cholera located in Bombay. Dr. Haffkine successfully developed a vaccine that was used throughout India reducing the incidence and mortality of the disease. He next turned his attention to developing a vaccine for plague. In India the pneumonic type of this disease was often epidemic. Caused by *Yersinia pestis*, the pneumonic form spreads by aerosol particles while the bubonic form is transmitted by fleas from infected rats. Plague was also endemic in China. In 1894, during an epidemic in Hong Kong, two foreign scientists came to work on the disease and independently identified the bacterial cause: Japanese pioneer immunologist, Shibasaburo Kitasato, M.D. (1852-1931), who together with Dr. Emil von Behring discovered the diphtheria antitoxin in 1890, and French immunologist, Alexandre Yersin, M.D., who discovered diphtheria toxin with Emil Roux, M.D. in 1888. In 1896, for the first time in 200 years, the plague, carried along the active trade route from Hong Kong, struck Bombay. Dr. Haffkine had been working on a plague vaccine since the isolation of the bacteria. He succeeded in protecting humans by inoculations of killed cultures. First he grew the cultures for purity then sterilized them using heat. To further assure the bacteria was dead he added carbolic acid to the solution. In the tradition of the era on January 10, 1897, Dr. Haffkine inoculated himself to test the vaccine. Although the side effects were considerable, pain, fever, tenderness and malaise, the vaccine proved to be effective though not an absolute protection. He then released the vaccine for general testing at the height of the Bombay epidemic. Of the 8,142 people inoculated, only 18 developed the plague. Standardization, however, was a problem. Further tests and manufacturing methods made safer vaccine available. By 1940 over 40,000,000 doses of Haffkine Plague Vaccine had been distributed and it proved to be the best protection available in the first half of the century. Dr. Haffkine became an international celebrity with his treatment of cholera and plague and was consulted during many epidemics. In 1925 the research center was renamed The Haffkine Institute in his honor.

10
Syphilitic Facial Destruction & The Development of Modern Art
1912

Art experts now hypothesize that the original stimulus for modern art originated from the study of the facial destruction in syphilitics, not from primitive African masks. Dr. William Rubin, Curator of Modern Art at New York's Museum of Modern Art, presented his thesis in the *Studies in Modern Art 3, Les Demoiselles d'Avignon.* This classic painting, considered the first work of modern art, was painted by Pablo Picasso and completed in 1907. It depicts the progressive degeneration of a woman until her face is a misshapen angular distortion. Picasso was a paradox. He was a misogynist yet a womanizer; frequented brothels yet deeply feared venereal disease, a disease he had contracted and had cured. Rubin documents not only Picasso's fascination with brothels but, more importantly, the fact that Picasso had permission from the French and Spanish governments to visit the syphilitic lazarettos. Among the strongest evidence Rubin presents on the validity of the assumption that Picasso did not base his work on the African mask is that the Mbuya (sickness) mask from Pende, Zaire, which is visually similar to Picasso's rendition, had been found not to be available in Europe until 1917. Even more fascinating is that this mask depicts the African form of syphilis, yaws. Rubin states, "As Picasso plumbed his unconsciousness and searched his memory for the most terrifying faces he had ever seen, he certainly must have recalled the ravages and distorted heads of some congenital syphilitics he had seen at Saint-Lazare. Such syphilitics are, in fact, the models for the Pende masks." In order to prove his thesis Rubin searched for examples of the extremes of facial destruction Picasso would have seen. He was unsuccessful until he discovered the images in The Burns Archive. In 1994 a series of these photographs were published in Rubin's book. Rubin noted "Dr. Burns, the author of one of the rare attempts to frustrate loss and destruction of this medical heritage…he has taken a lead in salvaging…in this case images, of syphilitic deformations that have disappeared from pathology." Photographs of horrendous deformities are usually the first images discarded from medical collections as conventions change. There can be no doubt that the most horrendous facial deformities caused by disease are those of congenital syphilis. Leprosy and cancer may destroy the face but the destruction usually occurs in an older person. Those with a severe form of congenital syphilis usually die as children though it is possible, even in the late stage of congenital syphilis, as seen here, they can live until their thirties. Fortunately these patients cannot see their faces, hear others or smell the odiferous, fetid oral and nasal area infections. They are blind, deaf and have lost their olfactory glands. The raw open wound ultimately becomes infected with deadly pathogens and the patient usually dies from pneumonia or other respiratory tract infection.

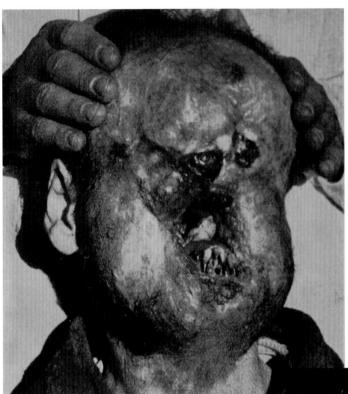
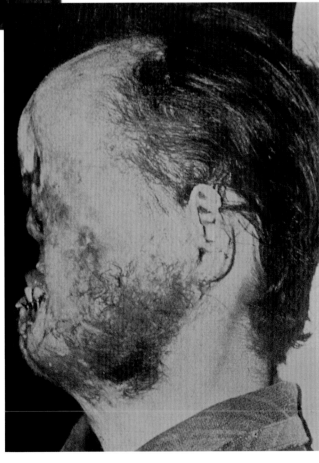

11
LISTENING TO CHEST SOUNDS WITH THE EAR TO THE BACK
GERMANY
1916

The stethoscope, the first of modern medicine's necessary tools, was invented in 1818 by Rene Laënnec, M.D. (1781-1826). It revolutionized diagnosis and treatment of chest disease and sparked the rise of the French School of Medicine. Because heart and lung sounds were magnified Laënnec was able to establish a disease classification system based on the various tones detected. The use of this instrument also enabled the physician to remain at a safer distance from an infected patient. In this photograph, taken in the throes of World War I, a military physician places his head on a patient's back to listen to lung sounds. Prior to the Laënnec stethoscope chest sounds were heard by simply placing the ear against the chest or back. As a quick screening method it worked for centuries. European physicians used this method and Laënnec's simple tube stethoscope, well into the 1940s. An American physician, Dr. Cammann of New York, invented the binaural stethoscope in 1855. This advanced instrument was largely ignored by European physicians, perhaps because of its American origin. In the nineteenth century it was the American who traveled to Europe to learn the latest techniques. European physicians only came to the United States as invited dignitaries. It was not until after World War I that America became the acknowledged world leader in medical research, technique and education.

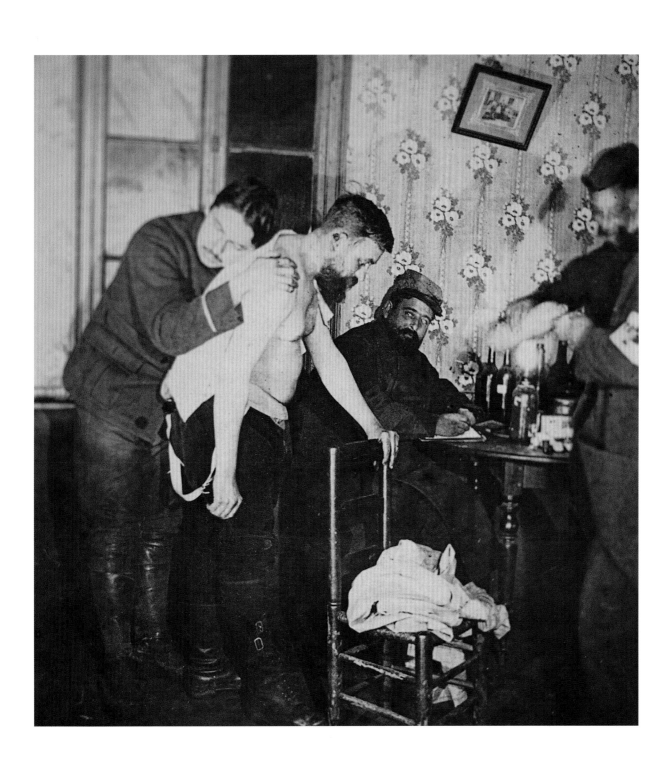

12
Inhalation and Massage therapy for Respiratory Tract Disease
1910

Until the development of specific medications a number of electrical, mechanical and inhalation therapies were used to treat respiratory tract disease. These were much safer treatments than bloodletting, blistering, purging and the use of mercury, opium and other powerful remedies. Electrical vibrators for chest and sinus massage were an early twentieth century innovation. Massage helped relieve congestion and improved circulation. The electrical device was efficient, inexpensive and allowed easy home use. By the second decade of the twentieth century electricity was available in most areas of the United States. Inhalation of a wide range of substances improved ventilation, cleared nasal and bronchial passages. Massage and inhalation therapies have remained an adjunct to modern therapy.

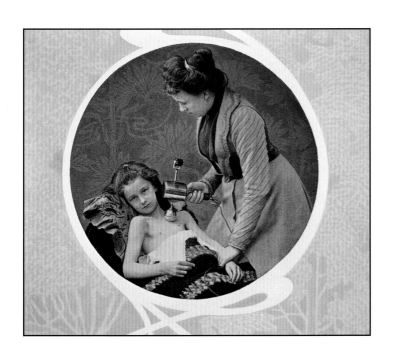

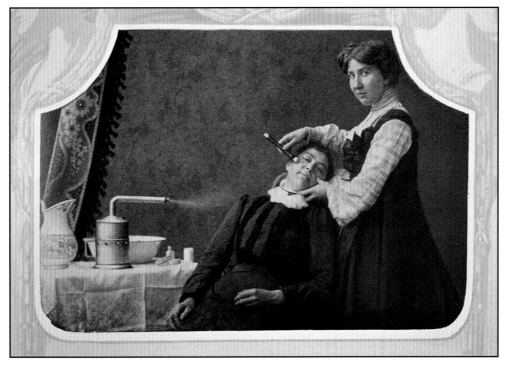

13
Cadaver Lung Injected with Bismuth
Stereoview
1915

In 1915, Kennon Dunham, M.D. published *Stereoroentgenography: Pulmonary Tuberculosis*. In this work Dunham produced a series of 41 stereoviews of x-rays of the lung in normal and diseased states as well as x-rays from cadavers. Dr. Dunham states, "I know that the X-ray has its limitations. What I bring you is my own creative work towards the diagnosis of tuberculosis. What I have done is adopt a practical technique (stereoroentgenography) to replace individual judgments with scientific accuracy." Dr. Dunham notes physical examination of patients by physicians with instruments such as the stethoscope are variable. The x-ray examination creates a permanent record that can be reviewed and interpreted by various physicians.

Stereoroentgenographs create a three dimensional view. Dr. Dunham's cadaver lung preparations were injected with bismuth to demonstrate various aspects of lung and chest anatomy. This technique also generated an image that was more easily read. This is a view (one half of the stereoview) of a lung removed from the body with "bismuth injected into the trachea, a (rubber) tube filled with 'shot' passing through the descending vena cava has been brought out of the heart, and the aorta can be seen passing over the left bronchus. There has been an unsuccessful effort made to inject one of the arteries of the upper right and one of the arteries of the upper left. This stereo has been produced to show the upper, middle and lower divisions on the right, with their respective trunks, and the upper and lower divisions of the left side, and to again call attention to where the bronchi are given off, because this is most important when trying to make out the location of the various lobes on these stereoscopic plates." Various forms of Dr. Dunham's stereoroentgenographs remain a part of modern radiological diagnosis.

14
'DON'T SPIT'
PUBLIC HEALTH MEASURES HELP CONQUER RESPIRATORY DISEASE
CINCINNATI, OH
1916

In the first decades of the twentieth century, sanitary laws and major public education efforts resulted in a dramatic decline of tuberculosis cases. In 1916, in Cincinnati, The Anti-Tuberculosis League enlisted a small army of boys who joyously flourished stencils and white paint throughout the city painting "Don't Spit on the Side Walk" on everything. Spitting was a very common practice even home spittoons were familiar objects. Tuberculosis caused constant phlegm in the throat and lungs and people needed to expectorate. With the recognition in the 1880s of the contagious nature of tuberculosis, control of body discharges became an important public health concern and new laws, legal enforcement and fines were enacted. In Cincinnati, as the signs were being painted, "The Health Department fined sixty persons for violating the regulations against spitting, fining each one a dollar and costs." This photograph appeared in *The Survey*, June 17, 1916. Along with the article on spitting appeared another article announcing the Rockefeller Foundation had given Johns Hopkins Hospital funds to establish a new institute of public health, under the auspices of the International Health Board. The hospital announced, "Henceforth 'Public Health' will be considered a 'profession' and that specialized study in preventive medicine and public health work will be accorded part of the study. The school states this new institute confirms the special worth of public health officers and gives a final denial to the saying 'already obsolescent' that anyone can do public health work."

15

TRACHEOSTOMY OPERATION, PERFORMED BY A RUSSIAN MILITARY SURGEON

POLAND
1916

As the Russian surgeon steadies his knife two assistants steady the patient. The doctor is about to perform a tracheostomy with the patient sitting up. During wartime procedures were often performed in unorthodox ways. The use of the tracheostomy, a life saving procedure, was not routinely employed until the work of Pierre Bretonneau, M.D. (1778-1862). In 1826, he reported saving the life of a child suffering from diphtheria by performing this operation. The procedure became a standard treatment for this disease. Bretonneau had also named this disease "diphtheria" when he found that croup, malignant angina, and scorbutic gangrene of the gums were all names for the same disease. Though the mortality rate of the procedure was calculated at about 75% this was probably incorrect as the seriousness of the underlying disease as a cause of death was not taken into consideration. The chief indications for the operation were mechanical airway obstruction and obstruction caused by accumulations of secretions not readily removed. In the first half of the twentieth century polio and diphtheria were the main causes of these conditions. Today 'prophylactic' tracheostomy is used in surgery to keep an open airway for those with low pulmonary reserve and/or a wet tracheobronchial tree that could not be easily cleared. The balloon cuffed endotracheal tube has eliminated the need for some tracheostomies.

One of the legends was in the treatment of President George Washington. The sixty-seven year old Washington on December 14, 1799 had a "septic sore throat" and was struggling for breath. One young physician in attendance, Elisha C. Dick, advised tracheostomy, a radical rarely performed procedure at the time. The elder physicians opted instead to use the popular "heroic therapy" regime of frequent bleeding, purging, blistering and vomiting. By morning Washington was dead.

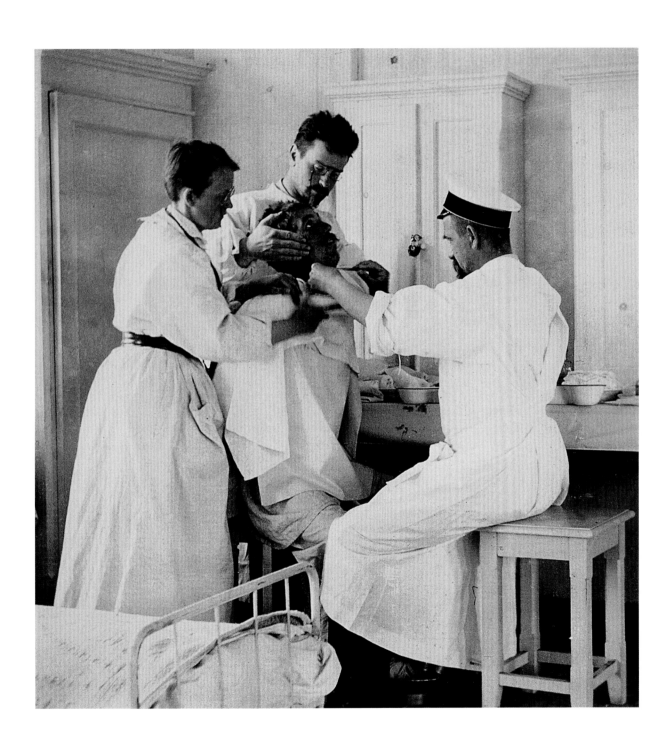

16
Hydrotherapy Compress Wrap for Lung Diseases
First Stage of Scotch Compress (Kreuz-Binde)
1914

Several types of physical therapies were used in the treatment of lung disease in the pre-antibiotic era. Among these were the use of fresh air and ventilation, as well as the application of simple applied dressings saturated with various chemicals. One of the most popular applications was hydrotherapy. These more elaborate water treatments were used by general physicians and specialist physician hydrotherapists. Physicians would often refer chronic lung disease patients to these specialists. Hydrotherapists offered a wide range of special cold or warm water dressing, often administered with frequent changes so the desired therapeutic temperature was maintained. The 'Cross' bandage, also called cloak dressing or Scotch compress consists of wrapping two sections of cold, well moistened, linen sheets, each three yards long and folded to twelve inches wide, entirely around the chest area in a prescribed manner. Note how in the first stage the sheet is being brought over the shoulder before wrapping the back. The advantage of this dressing was the complete coverage of the lung area providing maximum pleural and lung therapy. These chest dressings were applied cold. The treatment facilitated coughing, expectoration, relieved dyspnea and helped breathing. In some cases when the patient slept in the bandage overnight layers of oiled silk were laced between the wet cloth and the blankets to retain the temperature. The treatment varied with the seriousness of the condition. At times ice bags were used to lower the temperature. In edema and abscess of the lung hot thoracic compress dressings were applied at about 125 degrees. The hot compresses were also used for their sedative effect in angina pectoris and painful lung conditions such as pleuritis. The use of water as a remedial agent is probably as old as mankind and still has an important place in modern medicine, even lung disease. The most modern style hydrotherapy in lung disease involves being placed in an oxygen pressure chamber having one lung ventilated while the other lung is repeatedly filled with water solution to remove precipitated deposits.

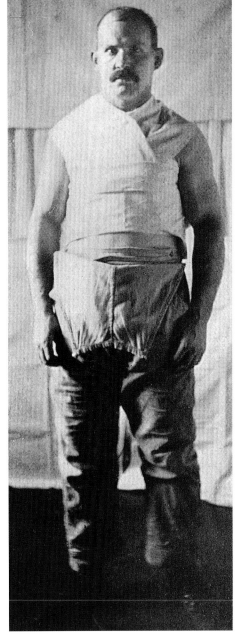

17
NON-PULMONARY COMPLICATIONS OF TUBERCULOSIS
FRANCE
1910

Modern physicians usually think of tuberculosis as a respiratory tract disease. However, in the tuberculin era when the disease was widespread, all areas of the body were attacked. The infection often created sinuses and fistulas that oozed pus. These draining wounds not only spread the disease further but created a nidus for frequently fatal secondary infection. Children and adults often lived with major parts of their bodies riddled with sinuses.

The treatment of tuberculosis was difficult and frustrating, with the patients suffering terrible pain and disability. One of the most devastating complications, especially in children, was osteomyelitis with abscesses and sinus formation. As this infection advanced it dissolved joints and bones grotesquely contorting the limbs. When it broke through to the skin, the joints and infected bones simply became riddled with holes that drained tuberculin pus. The condition would continue to progress until either the joint or bone was no longer recognizable or the patient died. The child in the top photograph was a ten-year-old boy who had lost nine pieces of his skull through the sinuses. In cases such as his, death occurred when the lesion penetrated the meninges or grew into the brain. All the patients in these photographs were treated at a center for sun therapy. By the end of two or more years of aggressive debridement with antiseptic agents and prolonged exposure in a high altitude to fresh air and sunshine, the wounds healed. Swiss physicians, Bernhard and Rollier, were the leading promoters of sun therapy. In 1903, August Rollier (1874-1954) introduced ultra-violet light and Alpine sunlight, at a high altitude, in the treatment of 'surgical tuberculosis'. He established a 'sun cure' center high in the Alps. By the second decade of the twentieth century "solar therapy" at special sanitariums was considered the best treatment for these indolent surgical cases. Some surgeons, such as Chicago's Emil Beck, were injecting the sinuses with bismuth under x-ray guidance. This treatment, a time consuming process involving skilled surgeons and radiologists, was repeated over several months. The sun cure relied more on cleanliness, nutrition and skilled nursing than physician intervention.

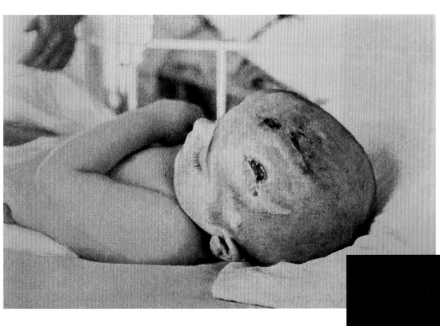

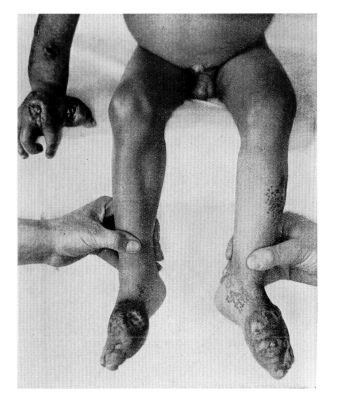

18
Heliotherapy:
An Important Aid to Control of Tuberculosis & Respiratory Disease
France
1914

Heliotherapy became an important adjunct in the control of tuberculosis and other respiratory tract diseases. The fresh air and rest cure of Edward Trudeau, M.D. (1848-1915), formulated in the last decades of the nineteenth century, resulted in the establishment of tuberculin sanitarium's emphasizing outdoor life. The next step was the idea that the sun had an important healing power. This concept actually dated back a millennium. In 1913, August Bernhard, after ten years of research, published, *Die Heliotherapie der Tuberculoses*. He firmly established the necessity for the use of sunlight in treating the disease. Heliotherapy is defined as "the systematic exposure of the nude body to the rays of the sun for treatment purposes." The advantage of the sun was said to be the blending of all ranges of radiation. The success of sun cures promoted the routine preventive use of 'sun exposure'. A portion of the school day in the early part of the century, continuing into the 1930s, was taking school children outdoors and exposing them to the sun, as seen here. Educating the public to the importance of the outdoors, sun exposure and health activity soon created a cult of sun exposure enthusiasts. The impetus for nudist camps emerged for this search for sun and health. The effort to get the public into the sun did generate an outdoor, health conscious society but it has also spawn a population with a soaring rate of skin cancer. Heliotherapy remained an important part of treatment for tuberculosis well into the 1940s. Dr. Richard Kovacs' 1942 treatise on *Electrotherapy and Light Therapy*, notes that heliotherapy is actively used for treating "pulmonary and non-pulmonary or surgical tuberculosis." In a survey of 602 tuberculosis institutions in the United States, 440 reported facilities for heliotherapy and recorded a total of 1,589,720 treatments.

19
Nobel Laureate, Dr. Neils Finsen and Phototherapy
Artificial 'Sun' Therapy
New York City
1918

It was the pioneer work of Danish physician, Neils Ryberg Finsen, M.D. (1860-1904) in light therapy that set other minds working to develop a wide range of light treatment modalities from heliotherapy to the sun lamp. In 1893, in Copenhagen, he began his experiments showing ultraviolet rays either stimulated growth or killed bacteria in lower organisms. In further research he studied the effect of light on living organisms and by 1896, had created the field of "phototherapy." Finsen was able to demonstrate that invisible ultraviolet light had therapeutic value. In 1899, he published his landmark text on the field, *La Phototherapie...Traitement du lupus vulgaire par des rayons chiminiques concentres* which demonstrated lupus could be successfully treated with light. For his discoveries Dr. Finsen received the Nobel Prize in Medicine in 1903. He died at the age of 44 a year after receiving his Nobel though his personal work lived on through his development of a burn technique treatment for lupus erythematosus, acne and eczema. A specialized machine was constructed for this treatment. It was seven-feet tall and contained four ports that made it possible to treat up to four patients at a time. The light was so concentrated it would deliver a burn about the size of a dime. This delivery system became a standard in European hospitals and clinics. Lupus erythematosus, acne and eczema were all treated with the Finsen burn technique.

The success of heliotherapy and other light techniques used in the treatment of tuberculosis led scientists to find a reliable source of artificial light. Not all climates or geographic areas offered sufficient conditions for sun exposure and city life had major obstacles. Scientists found that different wave lengths of light had different healing and therapeutic properties. The current, voltage, reflectors, applicators and a host of other modalities affected the intensity and delivery. By the 1920s a variety of light lamps were available. Special mercury vapor solariums were built where patients were treated by the dozen. All patients and personnel were required to wear protective glasses for this treatment although this baby is getting a sun lamp treatment and wears no glasses. Light therapy in various forms has remained an important modality for the treatment of a variety of conditions.

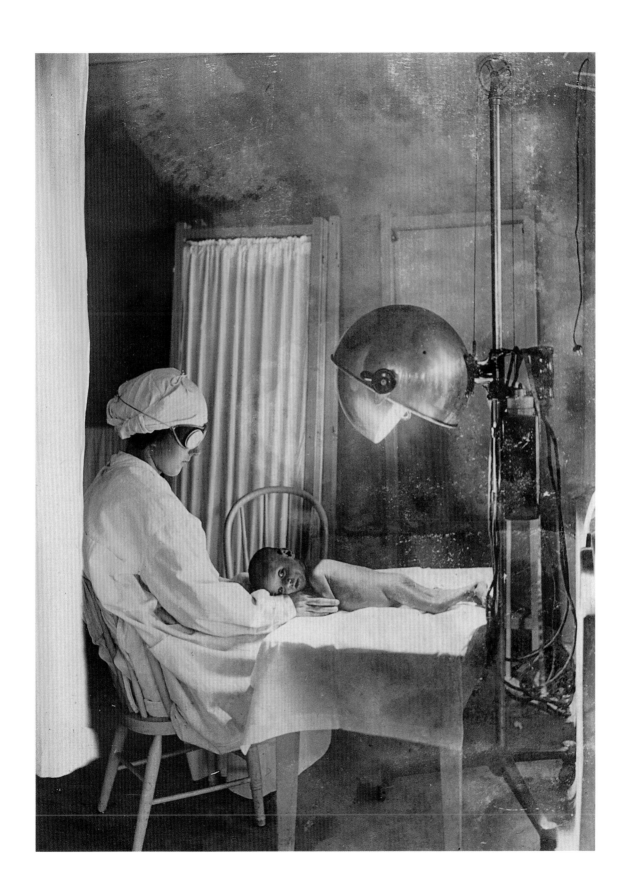

20
X-Ray of Catheter in Sinus

CIRCA 1919

Once developed the chest film became the standard screening device for tuberculosis. In World War I all servicemen and potential inductees were x-rayed. Use of the x-ray enabled the extent of tuberculosis in America to be calculated. During the first two decades of the twentieth century x-ray technology was rapidly improved with better bulbs, screens, and focusing devices, contrast media, catheters and probes. Two of the greatest advances, both in 1913, were William Coolidge's invention of the high vacuum, hot cathode, tungsten target x-ray tube and Dr. Gustave Buckey's collimating grid diaphragm. Buckey's device allowed clearer definition of internal structure and better control of the x-ray beam. Hollis Potter then improved the device further by inventing a method that made the grid movable and created an even clearer picture. This new process allowed the size of the image to be increased to the now familiar 14 by 17 inch standard film replacing the smaller glass plate negative. This x-ray of a metallic probe in a sinus is indicative of the advance in upper respiratory tract visualization provided by the Buckey-Potter grid and other technical refinements. During the second decade of the century progress in x-ray technology propelled surgery to elite status. With the x-ray surgeons could preoperatively and even operatively accurately evaluate a patient. Sinus cavities or diseased lung cavities could be seen and even outlined with contrast media. Surgery for all respiratory tract conditions became more accurate and commonplace. By 1920, the x-ray became an office based tool of the chest physician as necessary as the stethoscope.

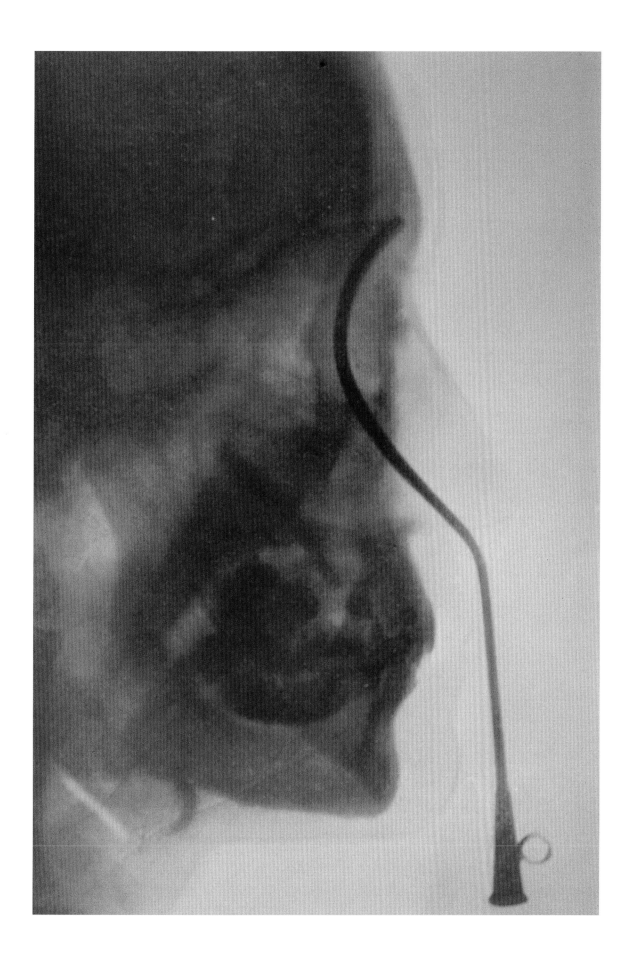

21
GERMAN WWI CAVALRY TROOPS AND THEIR HORSES IN GAS MASKS
FRANCE
1915

The respiratory system has become a focus of modern warfare for killing. Today the threat of gas warfare looms over us and is a major concern of both combatants and civilians. This photograph from World War I presents an anachronistic combat team if there ever was one. A German cavalry team is equipped with an ancient weapon, the lance, and wearing gas masks for protection against the most modern of weapons. Masks not only for themselves but for their horses! It is very difficult to imagine this mounted duo successfully surviving a gas barrage or charging through their own poison gas attack to lance their foe. But war has always had many surprises and incongruities. Gas warfare did not originate with the Germans, as generally thought. Actually the English first employed it in the Boer War in 1899. The Boer War forecast twentieth century events. The Boers used guerrilla warfare and the English responded with poison gas and a clever innovation, the civilian concentration camp. Typhoid and other diseases killed thousands of imprisoned old men, women and children. With their families locked up and dying, the Boers, though not defeated militarily, eventually surrendered. During this war the English used picric acid in artillery shells. When the projectile hit the ground, the shell released an explosive gas, lyddite. At the 1899 Hague Peace Conference a resolution was adopted, outlawing 'the use of projectiles the sole purpose of which is the diffusion of asphyxiating or deleterious gases.' Although the United States refused to sign the treaty, twenty-six other nations did. Great Britain did not sign until 1907. As with all legal agreements loopholes are the key to the future and the death of intent. In World War I, the autocratic German regime initially did not disobey the laws. They did not use artillery shells, but smoke pots and other devices from which released a lethal cloud of chlorine gas on April 22, 1915, at Ypres. The gas drifted over the French lines killing over 5,000 soldiers and injuring 10,000 more. Panic and confusion reigned as the French retreated. The Germans did not take advantage of the four-mile gap in the front line that would have led them to the English Channel and probable victory. They never did get to the Channel through France in World War I making it a prime objective of World War II.

Once gas was used it became a standard weapon in the war and every means including the outlawed artillery shells were used by both sides. Phosogene, mustard and chlorine gases were the main weapons but other gases were experimentally used. Gas caused an estimated 91,000 deaths and 1,200,000 casualties. Approximately 66,000 tons of toxic gases were used by the Germans; 58,000 tons by the Allies and 1,100 tons by the Americans. Mustard gas, though used late in the war, was delivered by an estimated 9,000,000 artillery shells, fired by both sides, and caused over 400,000 casualties. During the war the lethality of the gas steadily increased. The Americans introduced Lewisite, a more deadly gas that quickly blistered the skin and penetrated the body. America entered the war in 1917. The first gas attack against them was on April 25, 1918 when phosogene shells landed in their lines. Phosogene attacks the respiratory tract, killing by rapid edema and suffocation. America was prepared for the gas warfare, however, as they had entered late in the war and had time to prepare both defensively and offensively. On August 15, 1917, General Order 108 of the War Department created, for each army, one 'Gas and Flame Service Regiment'. It should be noted the flame thrower, a new and deadly weapon in World War I, used jellied gasoline. This substance was highly effective against tanks and in clearing out trenches and bunkers. By the end of the war the more refined gas mask and defensive measures became effective in protecting servicemen. The advantage of gas was lost as each side escalated and retaliated in turn. Poison gas use did not bring World War I to a close; it was another respiratory tract malady, influenza.

22
World War I German Soldier Posing with a Wax Model of His Original Wound

Lublin, Poland
1917

In trench warfare the most exposed part of the body is the head and neck. In World War I bullets and shrapnel extracted a heavy toll on both sides. Death was often the least cruel result, as tens of thousands had their faces mutilated in horrendous ways. Hundreds had such severe disfigurement that in order to appear in public the only recourse was to wear a mask. Others were so mutilated both eyes were lost and they spent the remainder of their lives in institutions until chronic infection claimed them. Many world-renowned sculptors donated years to making masks for the mutilated. After the war the facially mutilated formed special mutual aid societies. The devastation to the upper respiratory tract and face necessitated major reconstruction, innovative techniques and the creation of prosthesis. Maxillofacial surgery by dentists with facial reconstructions by plastic surgeons was one of the great surgical achievements of the war. Both surgical fields became firmly established by the end of the war. The work continued for years as the reconstructive procedures required multiple operations over long periods of time.

The medical achievements were not all on the Allied side. Germany and Austria had been leaders in late-nineteenth century-medicine and World War I gave them an ironic opportunity to fulfill their medical promise. At the Kriegszahnklink der IV Armee in Lublin, Poland, the latest surgical techniques allowed rehabilitation of the wounded. This work was documented in an unusual way. Moulange (wax modeling) had been a long standing European tradition. This technique was employed in making fully painted preoperative casts of wounds. This is a "before and after" photograph of a postoperative German soldier posed with the wax model of his original wound. The image is part of an album of 43 photographs that document the stages of treatment from initial surgery to the final result. Taken at the Lublin, Poland, German Fourth Army Dental-Maxillo Facial Unit these photographs are a remarkable document of the war. The album was preserved by a German Jewish dentist who served at the hospital during the war.

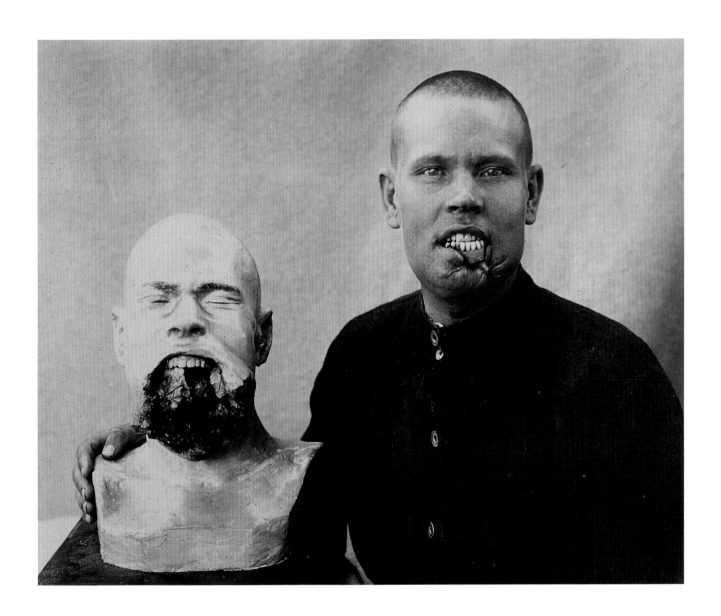

FT. FUNSTON INFLUENZA WARD - START OF THE WORLD WIDE PANDEMIC
KANSAS
SUMMER 1918

Acute respiratory infections are frequent and often fatal in both children and the elderly. The influenza pandemic of 1918-19 was noteworthy as it affected young adults causing their death by the millions and effecting an entire generation. Over 550,000 died of influenza in the United States, 10 times the number that died in WWI. The initial calculations estimated over 20 million had died worldwide. However, Asia, Africa and Latin America were under reported and millions more have to be added to the final toll. More people died in a six-month period than at any other time in the history of mankind. The exact nature of the killer virus was never unraveled. Some hypothesize it was a synergy of a virus with a bacterial agent that produced a deadly pneumonia while others hypothesize it was an unusually powerful pathological, antigenic, altered virus that produced inflammation with rapid edema that choked a patient to death. Curiously the disease was called the Spanish Flu not because the disease started there but because Spain was not a belligerent nation in WWI and did not have news censorship. This allowed the prevalence and devastation wrought by this disease to become a news item. In reality the disease probably started in the United States at Fort Funston, Kansas. This image of the first influenza ward of the pandemic dramatically illustrates the virulence of the disease.

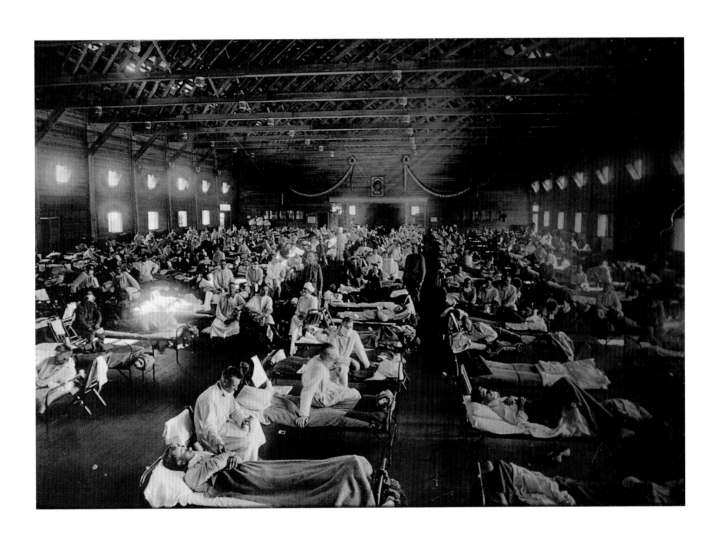

24
BASEBALL GAME DURING THE INFLUENZA PANDEMIC
FALL 1918

The public appreciation for the highly contagious nature of the influenza during the epidemic of 1918 is well documented in this photograph. The image is an icon of public health awareness and positive public response. Everyone in the photograph is wearing a protective face mask: the batter, catcher, umpire and all the spectators. Millions were ill with a viral pneumonia and the entire health care system was strained to the limit. Some morgues had to function with all their windows and doors open as the bodies were "stacked like cord-wood" and the stench overpowering. Soldiers trying to report to sickbay had to weave in and out of bodies blocking the entrances to the hospital. In many cities all public gathering, business and entertainment was forbidden. Citizens scrubbed the streets and sidewalks at night trying to prevent the spread of the disease. In the fall of 1918, in New York City alone 2,000 were dying each week, 500,000 were infected with the end total of 45,000 dead. The pandemic only lasted through the winter and disappeared as quickly as it had arrived.

25

AMERICAN WOMAN'S HOSPITAL NO.I

DRS. HORREL & BENTLY ON WARD ROUNDS
LUZANCY, FRANCE
CIRCA 1918

Before the United States entered World War I, American medical volunteers, medical institutions and social groups established and supported their own hospitals in Europe. When America entered the war numerous medical institutions continued this practice, however, their medical personnel were required to join the military service. There was similar community behavior during the Civil War when local groups joined together to form their own regiments. Many noted American surgeons volunteered for duty in this fashion serving not only as physicians but hospital directors as well. Some of the 'sister' hospitals were Boston's Peter Brent Brigham Hospital was Base Hospital # 5; Cleveland's Lakeside Hospital was Base Hospital # 4; Northwestern University was Base Hospital # 12; University of Pennsylvania was Base Hospital # 10, University of Washington was Base Hospital # 50. The eminent neurosurgeon, Harvey Cushing, M.D., was Medical Director of Harvard's Base Hospital # 5. In this photograph taken at American Woman's Hospital No. 1, Luzancy, France, two women physicians, Dr. M. Louise Horrell and Inez C. Bently are on ward rounds. Dr. Horrell holds a binaural stethoscope as she reads a patient's chart. This image was one of a series taken to promote the work of the hospital. After the war the hospital and its volunteers remained in Europe to serve the civilian population.

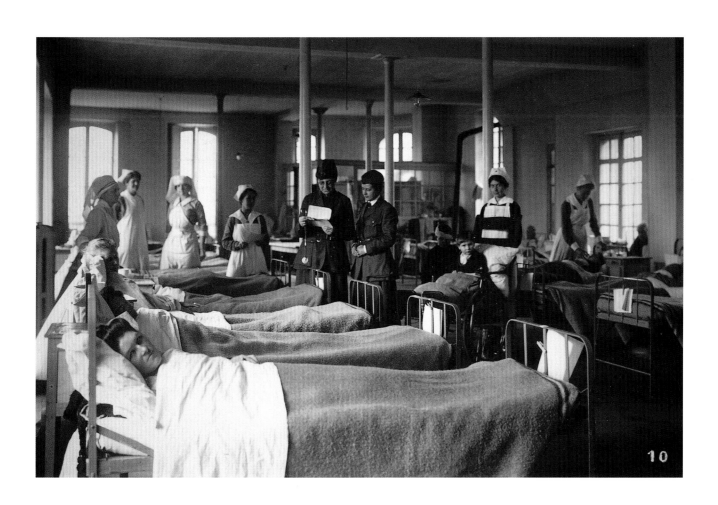

26
FEEDING TUBE INSERTED INTO WOUNDED WWI SOLDIER

US ARMY CAPTAIN HILL
MAXILLO-FACIAL WARD, WALTER REED US ARMY HOSPITAL
WASHINGTON, DC
1919

A massive jaw wound necessitates the insertion of a tube into the mouth to assist in feeding and secretion management of wounded US Army Captain Hill. Two years later after his wound healed Captain Hill proudly shows off his lower jaw prosthesis. These photographs are from Walter Reed US Army Hospital's maxillo facial surgery service. The photographs were taken by Walter Reed's Alice C. Becht and are part of a series used to illustrate the 1920 classic text on reconstructive surgery, Dr. Vilray Blair's *Surgery and Diseases of the Mouth*. American maxillofacial as well as reconstructive and plastic surgery became firmly established as a result of the treatment of hundreds of patients with massive facial wounds. By the 1920s with the techniques learned treating World War I wounded, the surgeons used the procedures in treating head and neck cancers.

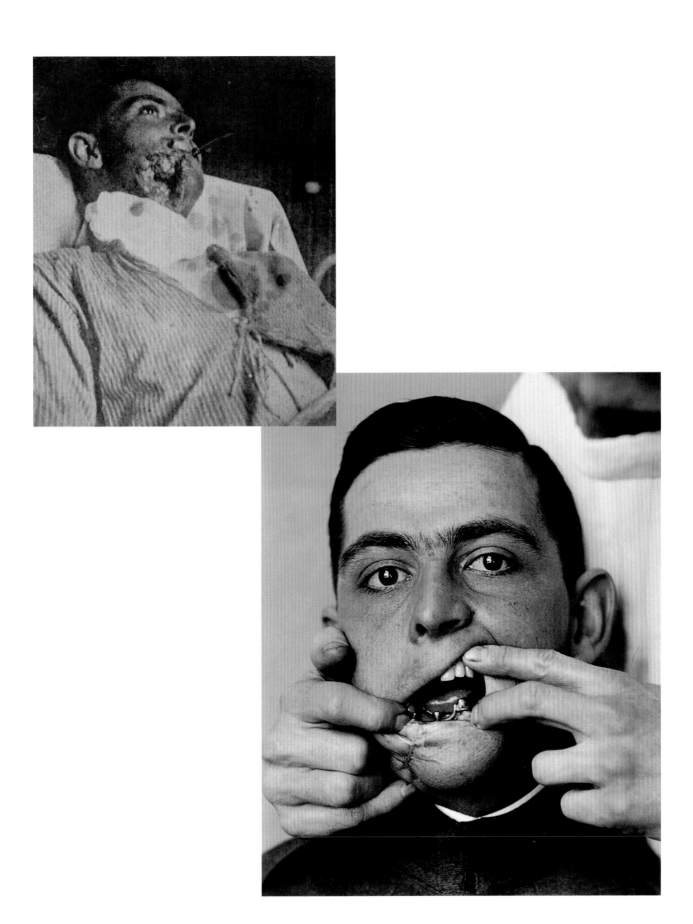

BIBLIOGRAPHY

Beck, Emil G., M.D., *Bismuth Paste Injections, Part First, Part Second, Stereo-Clinic.* Southworth Co.: Troy, NY, 1911.

Bernheim, Bertram M., M.D., *The Story of Johns Hopkins: Four Great Doctors and the Medical School they Created.* McGraw Hill Book Co.: New York, NY, 1948.

Bordley III, James, M.D., and A. Harvey McGehee, M.D. *Two Centuries of American Medicine: 1776-1976.* W.B. Saunders Co.: Philadelphia, Pennsylvania, 1976.

Brieger, Gert H. *Medical America in the Nineteenth Century: Readings from the Literature.* Johns Hopkins Press: Baltimore, Maryland, 1972.

Burdick, Gordon G., *X-Ray and High Frequency in Medicine. Physical Therapy.* Library Publishing Co.: Chicago, IL,1909.

Burns, Stanley B., M.D., and Richard Glenner, D.D.S. et al. *The American Dentist: A Pictorial History with a Presentation of Early Dental Photography in America.* Pictorial Histories Publishing Co.: Missoula, MT, 1990.

Burns, Stanley B., M.D., and Ira M. Rutkow, M.D. *American Surgery: An Illustrated History.* Lippincott-Raven Publishers: Philadelphia, PA, 1998.

Burns, Stanley B., M.D. *Early Medical Photography in America: 1839-1883.* The Burns Archive: New York, NY, 1983.

Burns, Stanley B., M.D., and Sherwin Nuland, M.D. et al. *The Face of Mercy: A Photographic History of Medicine at War.* Random House: New York, NY, 1993.

Burns, Stanley B., M.D., and Joel-Peter Witkin, et al. *Masterpieces of Medical Photography: Selections From The Burns Archive.* Twelvetrees Press: Pasadena, CA 1987.

Burns, Stanley B., M.D. *A Morning's Work: Medical Photographs from The Burns Archive & Collection 1843-1939.* Twin Palms Publishers: Santa Fe, New Mexico, 1998.

Burns, Stanley B., M.D., and Jacques Gasser, M.D. *Photographie et Médecine 1840-1880.* Insitut universitaire d'histoire de la santé publique: Lausanne, Switzerland, 1991.

Burns, Stanley B., M.D. and Elizabeth A. Burns. *Sleeping Beauty II: Grief, Bereavement and The Family in Medical Photography, American & European Traditions.* Burns Archive Press: New York, NY, 2002.

Clarke, Edward H., M.D. et al. *A Century of American Medicine: 1776-1876.* Burt Franklin: New York, NY, 1876.

Crowe, Samuel James, M.D., *Halstead of Johns Hopkins: The Man and His Men.* Charles C. Thomas, Publisher: Springfield, Il, 1957.

Cummins, S. Lyle, M.D. *Tuberculosis in History: From the 17th Century to our Times.* Bailliere, Tindall and Cox: London, 1949.

Daniel, Thomas M. and Frederick C. Robbins, Editors. *Polio.* University of Rochester Press: New York, 1997.

Davis, Loyal. *Fifty Years of Surgical Progress: 1905-1955.* Franklin H. Martin Memorial Foundation: Chicago, Illinois, 1955.

Dieffenbach, William H. *Hydrotherapy: A Brief Therapy of the Practical Value of Water in Disease for Students and Practicians of Medicine.* Rebman Co.: New York, NY, 1909.

Donahue, M. Patricia. *Nursing: The Finest Art.* Mosby: St. Louis, Missouri, 1996.

Duffy, John. *The Healers: A History of American Medicine.* University of Illinois Press: Urbana, Illinois, 1976.

Dubos, Rene and Jean. *The White Plague: Tuberculosis, Man and Society.* Little Brown and Company: Boston, MA, 1952.

Dunham, Kennon. *Stereoroentgenography Pulmonary Tuberculosis, Part First and Part Second, Stereo-Clinic.* Southworth Co.: Troy, NY, 1915.

Editors. *Who's Important in Medicine.* Institute for Research in Biography Inc.: New York, NY, 1945.

Fee, Elizabeth and Daniel M. Fox. *AIDS: The Burdens of History.* University of California Press: Berkeley, California, 1988.

Frizot, Michel. *The New History of Photograph.,* Könemann Verlagsgesellschaft mbH: Koln, Germany, 1998.

Fye, W. Bruce, M.D. *The Development of American Physiology: Scientific Medicine in the Nineteenth Century.* Johns Hopkins University Press: Baltimore, Maryland, 1987.

Garrison, Fielding H. M.D. *An Introduction to the History of Medicine. With Medical Chronology, Suggestions for Study and Bibliographic Data.* W.B. Saunders Co.: Philadelphia, Pennsylvania, 1913.

Glaser, Gabrielle. *Doctors Rethinking Treatments for Sick Sinuses.* The New York Times, Health & Fitness Section, Dec. 17, 2002, p. F6.

Gould, Tony. *A Summer Plague: Polio ad Its Survivors.* Yale University Press: New Haven and London, 1995.

Hendrickson, Robert. *More Cunning than Man: A Social History of Rats and Man.* Dorset Press: New York, NY, 1983.

Hersh, Seymour M. *Chemical and Biological Warfare: America's Hidden Arsenal.* Bobbs-Merrill Company: Indianapolis, 1968.

Hochberg, Lew, M.D. *Thoracic Surgery Before the 20th Century.* Vantage Press: New York, NY, 1960.

Hopkins, Donald R. *Princes and Peasants: Smallpox in History.* University of Chicago Press: Chicago and London, 1983.

Hurwitz, Alfred, M.D. and George Degenshein, M.D. *Milestones in Modern Surgery.* Hoeber-Harper, New York, NY, 1958.

Isselbacher, Kurt, M.D., et al. Eds. *Harrison's Principles of Internal Medicine, Thirteenth Edition.* McGraw Hill: Health Professionals Division, 1994.

Johnson, Stephen L. *The History of Cardiac Surgery: 1896-1955.* Johns Hopkins Press: Baltimore, Maryland, 1970.

Keen, William W., M.D. *Surgery; Its Principles and Practice, by Various Authors.* W.B. Saunders Co.: Philadelphia, PA, 1908.

Kelly, Howard and Walter Burrage. *Dictionary of American Medical Biography.* D. Appleton and Co.: New York, NY, 1928.

Kevles, Bettyann Holtzmann. *Naked to the Bone: Medical Imaging in the Twentieth Century.* Helix Books, Addison Wesley: Reading, Massachusetts, 1997.

Kiple, Kenneth F. *The Cambridge World History of Human Disease.* Cambridge University Press: New York, NY, 1993.

Kovacs, Richard, M.D. *Electrotherapy and Light Therapy: With Essentials of Hydrotherapy and Mechanotherapy.* Lea & Febiger: Philadelphia, 1942.

Leibowitz, J.O. *The History of Coronary Heart Disease.* Wellcome Institute of the History of Medicine: London, 1970.

Levinson, Abraham, M.D. *Pioneers of Pediatrics.* Froben Press: New York, NY, 1936.

Lopate, Carol. *Women in Medicine.* Johns Hopkins Press: Baltimore, Maryland, 1968.

Lyons, Albert S., M.D. and J.S. Petrucelli II, M.D. *Medicine: An Illustrated History.* Harry N. Abrams, Inc.: New York, 1978.

Bibliography

Margotta, Roberto. *The Story of Medicine*. Golden Press: New York, NY, 1967.

McHenry, Lawrence C. Jr., M.D. *Garrison's History of Neurology*. Charles C. Thomas: Springfield, IL, 1969.

McNeil, Donald G. *Combined Vaccine Gets F.D.A. Approval,* The New York Times, Dec. 17, 2002, p. A 33.

Morton, Leslie T. *A Medical Bibliography (Garrison and Morton): An Annotated Check-List of Texts Illustrating the History of Medicine*. Andre Deutsch, Morrison & Gibb, Ltd: London, 1970.

Packard, Francis R., M.D. *History of Medicine in the United States*. Hafner Press: New York, NY, 1973.

Parascandola, John, ed. *The History of Antibiotics: A Symposium*. American Institute of the History of Pharmacy: Madison, Wisconsin, 1980.

Puderbach P. *The Massage Operator*. Benedict Lust: Butler, New Jersey, 1925.

Reverby, Susan M. *Ordered to Care: The Dilemma of American Nurisng, 1850-1945*. Cambridge University Press, New York, NY, 1987.

Rice, Thurman B., M.D. *The Conquest of Disease*. Macmillian Co.: New York, NY, 1932.

Rothstein, William G. *American Physicians in the Nineteenth Century: From Sects to Science*. Johns Hopkins University Press: Baltimore, Maryland, 1972.

Rowntree, Leonard G. M.D. *Amid Masters of Twenthieth Century Medicine: A Panorama of Persons and Pictures*. Charles C. Thomas: Springfield, IL, 1958.

Rubin, William, et. al. *Les Demoiselles d'Avignon, tyudies in Modern Art 3, Museum of Modern Art*. Harry Abrams, Inc. Distributors: New York, NY, 1984.

Sarnecky, Mary T., DNSc. *A History of the US Army Nurse Corps*. University of Pennsylvania Press: Philadelphia, PA, 1999.

Schmidt, J.E., M.D. *Medical Discoveries: Who and When*. Charles C. Thomas: Springfield, IL, 1959.

Silverstein, Arthur M. *A History of Immunity*. Academic Press, Inc.: San Diego, 1989.

Tauber, Alfred and Chernyak. *Metchnikoff and the Origins of Immunology: From Metaphor to Theory*. Oxford University Press: New York, NY, 1991

Walker, Kenneth. *The Story of Medicine*. Oxford University Press: New York, 1955.

Wallace, Antony F. *The Progress of Plastic Surgery: An Introductory History*. Willem A. Meeuws: Oxford, England, 1982.

Wangensteen, Owen H., M.D., PhD. and Sarah D. Wangensteen. *The Rise of Surgery: From Empiric Craft to Scientific Discipline*. University of Minnesota Press: Minneapolis, MN, 1978.

Weir, Neil. *Otolaryngology: An Illustrated History*. Butterworths: London, England, 1990.

Wilson, David. *In Search of Penicillin*. Alfred A. Knopf: New York, 1976.

Winslow, Charles-Edward Amory. *The Conquest of Epidemic Disease: A Chapter in the History of Ideas*. University of Wisconsin Press: Madison, WI, 1943.

Worden, Gretchen. *Mütter Museum of the College of Physicians of Philadelphia*. Blast Books: New York, NY, 2002.

Wright, Jonathan, M.D., *A History of Laryngology and Rhinology*. Lea & Febiger: Philadelphia, PA, 1914.

Photographic Formats

The Evolution of Popular Photographic Processes 1850-1920

In 1851 the wet plate process became the dominant paper photographic process. This method used collodion as a base to hold the silver sensitive material. The solution was then spread over a glass plate. The plate was inserted into the camera, while still wet, exposed while wet, and then immediately developed while still wet. The photographer had to be both a plate maker and processor. Professional photographers produced almost all of the photographs in the 1851-1881 era.

In 1871, a physician, Richard Leach Maddox (1816-1902), changed the nature of photography with his discovery of the dry plate photographic process. A well-known British photomicrographer, Dr. Maddox contributed his images to books on microscopy. He discovered the long searched-for vehicle that had eluded thousands of researchers by using a gelatin bromide base as the sensitive medium instead of collodion. When this gelatin based silver solution dried it could be used at anytime. By 1878 the process was perfected to the degree that it was possible to take photographs in 1/25th of a second.

The dry plate freed the photographer from plate making and gave him greater mobility, as he didn't have to take the developing tanks and apparatus along. If he wished he could simply send the plates to one of the major photographic supply houses for developing and printing. The dry plate allowed thousands of amateurs to join the photographic ranks. In the mid 1880s we begin to see physicians taking private photographs with more intimate poses.

From 1889 forward, anyone could become a photographer. George Eastman introduced to the public a preloaded camera with flexible film capable of 100 photographs. The photographer shot the roll of film and sent the entire camera and film back to Kodak. The photographs were developed and printed, the camera was loaded with another 100 shots and everything was returned to the customer. By 1900, a simpler camera model was introduced named the 'Brownie.' Kodak used the motto "You Press the Button we do the Rest." Physicians could now easily take their own images of patients or their work places. These medical snapshots allowed personal images of the profession. In 1912, Kodak introduced the modern flat flexible 'film' in place of the glass dry plate. It was convenient, thin, lightweight, and unbreakable. It had an important impact in medicine as x-rays could now to be produced with faster exposures times and larger sizes. The 14 by 17 inch chest film became standard.

During the years 1908-1920 the photographic postcard became the most popular size used by the amateur photographer. Billions of these images were produced and sent through the mails. Many of the 'postcards' reproduced in this book are unique, personal images. While we think of the postcard as a commercial production, in this early era both amateur and professional photographers produced them.

The medical photographs in this volume are all originally paper based silver prints or photogravures from silver prints. Because of the relative uniformity of the photographic process the exact type of print has not been identified on each individual photograph in order to concentrate on the subject matter.

DEDICATION

Many physicians who treated and investigated tuberculosis contracted and succumbed to the disease. The list is voluminous and contains many medical luminaries including Dr. Rene Laënnec, inventor of the stethoscope. The sacrifices made by physicians for the advancement of medicine are legendary. Those who died or were mutilated in the development of radiology are well known but the specialists and conquerors fighting tuberculosis have never been fully appreciated. To these selfless individuals, I dedicate this work.

ACKNOWLEDGEMENTS

First and foremost I would like to thank my family who are integral parts of the Burns Archive. They have assisted me, tirelessly, in preparing this historic compilation. My wife, Sara, helps with collecting, cataloging and archiving the collection, and directs the stock photography use of the material. More importantly she serves as my sounding board and editor helping to clarify my ideas. My daughter, Elizabeth, designed and directed the entire production of these volumes from their conception to the final product.

I am most grateful to Saul G. Hornik, MS, RPh, medical marketing consultant. It was his enthusiasm and recognition of the educational importance of my medical, photographic collection, together with his tireless work that made this publication a reality. I give my thanks to Eric Malter, President of MD Communications, for his support of this project. I also wish to express my sincere appreciation to Christopher J. Carney, Director of Training Services of GlaxoSmithKline for understanding the educational value in using the visual history of the past as a foundation for the future.

STANLEY B. BURNS, M.D., F.A.C.S.

Stanley B. Burns, M.D., F.A.C.S., a practicing New York City ophthalmic surgeon, is also an internationally distinguished photographic historian, author, curator and collector. His collection, started in 1975, is considered to be the most comprehensive private, early, historic photograph collection in the world. Contained within this archive of over 700,000 vintage prints is the finest and most comprehensive compilation of early medical photographs consisting of 50,000 images taken between 1840 and 1940. These medical photographs have been showcased in countless publications and films, museum exhibitions. France's Channel Plus prepared a documentary on his work as part of the *Great Collectors of the World Series*. Dr. Burns has been an active medical historian since 1970. From 1979-81, he was President of the Medical Archivist of New York State. He has been a member of the medical history departments of The Albert Einstein College of Medicine and The State University of New York, Medical College at Stony Brook; Curator of photographic archives at both The Israeli Institute on The History of Medicine (1978-1993) and The Museum of The Foundation of The American Academy of Ophthalmology. Currently, he is a contributing editor for five specialty medical journals. The Burns Archive, his stock photography and publishing entity, is a valuable photographic resource for both researchers and the media. Using his unique collection Dr. Burns has written ten award-winning photo-history books, hundreds of articles and curated dozens of exhibitions. His film company, Black Mirror Films, produced *Death in America*, a documentary on the history of death practices in America. He is currently working on several medical exhibitions and books, as well as photographic history books on criminology, Judaica, Germans in WW II and African Americans. He can be reached through his web site www.burnsarchive.com.

OTHER BOOKS

Sleeping Beauty II: Grief, Bereavement and The Family in Memorial Photography, American & European Traditions

A Mornings Work: Medical Photographs from The Burns Archive & Collection, 1843-1939

Forgotten Marriage: The Painted Tintype & The Decorative Frame 1860-1910, A Lost Chapter in American Portraiture

American Surgery: An Illustrated History
co-author: Ira M. Rutkow, M.D.

Harm's Way: Lust & Madness, Murder & Mayhem
co-authors: Joel-Peter Witkin, et al

The Face of Mercy: A Photographic History of Medicine at War
co-authors: Matthew Naythons, M.D. and Sherwin Nuland, M.D.

Photographie et Médecine 1840-1880
co-author: Jacques Gasser, M.D.

Sleeping Beauty: Memorial Photography in America

The American Dentist: A Pictorial History
co-authors: Richard Glenner, D.D.S. and Audrey Davis, PhD.

Masterpieces of Medical Photography: Selections From The Burns Archive
co-author: Joel-Peter Witkin

Early Medical Photography in America: 1839-1883

THE BURNS ARCHIVE PRESS
140 EAST 38TH STREET • NEW YORK, N.Y. 10016
TEL: 212-889-1938 • FAX: 212-481-9113 • WWW.BURNSARCHIVE.COM